NATURE'S *Secret* SUPERHEROES

ELEPHANTS

THE LEAGUE OF EXTRAORDINARY GIANTS

Stephanie Feldstein

Cherry Lake Press

Published in the United States of America by Cherry Lake Publishing Group
Ann Arbor, Michigan
www.cherrylakepublishing.com

Reading Adviser: Beth Walker Gambro, MS, Ed., Reading Consultant, Yorkville, IL
Book Designer: Dan Winter

Photo Credits: © David Steele/Shutterstock, cover, title page; © Wijnand vT/Shutterstock, 5; © Fabinho Silva/Shutterstock, 6; © RuqayaMai/Shutterstock, 7; © mst rabia akter/Shutterstock, 7; © Robert Harding Video/Shutterstock, 9; © AlexUm5/Shutterstock, 11; © MRAUT_photos/shutterstock, 12; © Vibe Images/Shutterstock, 13; © mst rabia akter/Shutterstock, 14; © imageBROKER.com/Shutterstock, 15; © Sunny Photoss/Shutterstock, 17; © Roger de la Harpe/Shutterstock, 18; © GS de Carvalho/Shutterstock, 20; © Gaston Piccinetti/Shutterstock, 21; © mst rabia akter/Shutterstock, 22; Kris Krüg, CC BY-SA 2.0 via Wikimedia Commons, 23; © RMFerreira/Shutterstock, 25; © Sebastian Castelier/Shutterstock, 26; © mst rabia akter/Shutterstock, 28; © Jishith Jayaram/Shutterstock, 29; © Independent birds/Shutterstock, 30 © mst rabia akter/Shutterstock, 30;

Cherry Lake Press is an imprint of Cherry Lake Publishing Group.

Library of Congress Cataloging-in-Publication Data

Names: Feldstein, Stephanie author
Title: Elephants : the league of extraordinary giants / written by Stephanie Feldstein.
Description: Ann Arbor, Michigan : Cherry Lake Publishing, [2025] | Series: Nature's (secret) superheroes | Audience term: Juvenile | Audience: Grades 4-6 Cherry Lake Publishing | Summary: This title from our Nature's (Secret) Superheroes series delves into the powerful effect elephants have on their environment. Learn how these underappreciated gentle giants work as environmental superheroes to keep their ecosystems going strong-- Provided by publisher.
Identifiers: LCCN 2024035981 | ISBN 9781668958971 pdf | ISBN 9781668958100 ebook | ISBN 9781668957233 paperback | ISBN 9781668956380 hardcover
Subjects: LCSH: Elephants--Juvenile literature.
Classification: LCC QL737.P98 F4425 2024 | DDC 599.67--dc23/eng/20240826
LC record available at https://lccn.loc.gov/2024035981
Cherry Lake Publishing Group would like to acknowledge the work of the Partnership for 21st Century Learning, a Network of Battelle for Kids. Please visit Battelle for Kids online for more information.

Printed in the United States of America

Note from publisher: Websites change regularly, and their future contents are outside of our control. Supervise children when conducting any recommended online searches for extended learning opportunities.

TABLE OF CONTENTS

ADVENTURES IN THE SAVANNA

Some of the most beloved animals on the planet live in Africa's savannas. A savanna is a flat, grassy landscape. Trees are spread out with open spaces in between. Giraffes, zebras, and rhinoceroses live there. Lions, cheetahs, and hyenas live in savannas, too. And so do, of course, the biggest animals of the savanna, elephants.

Savannas have a wet season and a dry season. During the summer rains, it's easy for the animals to find water. The rest of the year, the land is dry. Animals rely on watering holes to survive. A watering hole is a place where the ground dips. It fills with water to form a small lake. Hundreds of kinds of animals gather there to drink. Elephants love the water. They play in it. They spray themselves with water to keep cool.

Elephants can create their own watering holes. They dig with their feet, trunks, and tusks. They dig until they reach water deep underground. The water fills the hole. Then other animals can drink there, too.

Elephants do have a good memory! They remember the locations of watering holes up to 31 miles (50 kilometers) away.

Ecosystems are places where plants, animals, and the environment rely on each other. They help produce food, shelter, water, and clean air. Every kind of **native species** is important. If one disappears, it can harm the whole ecosystem. But some species are secret superheroes. They play an especially big part in shaping the web of life.

Elephants don't exactly live undercover. They have a big presence wherever they go. But many people don't realize that elephants have superpowers. They make watering holes and plant trees. They shape their ecosystem so lots of other species can survive.

It's not easy being a superhero. Elephants have been hunted close to **extinction** for their tusks. Extinction is when all of one kind of plant or animal die. Elephants are more protected than they used to be. But they're still being killed in alarming numbers. These superheroes need our help so they can keep protecting the planet.

ELEPHANT TALES

Elephants show up in a lot of children's stories and movies. One of the most famous elephant characters is Dumbo. In the Disney movie, Dumbo was a baby elephant kept captive in the circus. Real baby elephants are sometimes taken from their families and trained to perform tricks in circuses. But places around the world are making it illegal to use elephants in circuses.

Everyone made fun of Dumbo for his big ears. But then he learned to use his ears to fly. Real elephants can't fly. But they do have giant ears. And those ears do more than just hear.

African elephant ears can be up to 6 feet (1.8 meters) long and 5 feet (1.5 m) wide!

African elephants have the biggest ears of any animal on the planet. They use them like fans when it's hot. Their ears also have lots of blood vessels. Their blood cools down as it moves through the thin flaps of their ears. That helps the rest of their body stay cool. They can even control their blood flow. They can make more blood move through their ears when they need to cool down.

CHAPTER 1

ORIGIN STORY

Elephants are the largest animals that live on land. Whales are the only animals that are bigger. The three kinds of elephants are African savanna elephants, African forest elephants, and Asian elephants. African savanna elephants are the biggest. They can be 30 feet (9.1 m) long from trunk to tail. They can weigh as much as 15,000 pounds (6,804 kilograms).

An elephant trunk is more than just a long nose. Of course, elephants use their trunks to breathe and smell things. Swimming elephants can stick their trunks out of the water like a snorkel to breathe. But they also use trunks like we use our hands and arms.

Elephants can pluck leaves from trees with the end of their trunks. They can pull up grasses. Then they use their trunks to bring the food to their mouths. They use their trunks to drink water. But it's not like a straw. It's more like a glass that scoops up water and carries it to their mouth.

Elephants have up to 40,000 muscles in their trunks.

Elephants also use their trunks to comfort each other. Just like you might put an arm around a friend's shoulder to make them feel better, elephants put their trunks around their friends.

Elephants live almost as long as people. Asian elephants can live to be 74 years old. African elephants can live up to 80 years.

Elephant families are called herds. Herds are mostly female elephants and their children living together. Males leave the herd when they're about 14 years old. They form their own groups. Some males roam on their own. But they often still visit their relatives.

Herds can be made up of up to 70 elephants.

DID YOU KNOW?

Elephants can stand up on their back legs to reach tall tree branches. But they can't jump. That's because all the bones in their legs point down. It's like they're standing on the tips of their toes. This helps support their weight. But it also means they can't spring into the air.

Elephants communicate in many different ways. They use body language, sounds, scents, and vibrations.

Depending on the species, an elephant can live up to 80 years. That's many years of knowledge to pass on to the next generation.

Elephants are very social. They can be friends with other elephant families. They recognize each other's faces. Their herds are always changing. An elephant might sleep at night with its relatives. Then it might hang out with a different herd during the day. It's like spending time with one group of people at school and another at home.

The **matriarch** is the head of the family. She's the oldest female in the herd. She remembers the best places to find food and water. Sometimes the herds have to travel long distances to get there. But she always remembers. She passes her knowledge to the younger elephants so they'll remember for the next generation.

FAMILY TREE

Two of the elephant's closest relatives live in the sea. Dugongs and manatees are sometimes called sea cows. They look a bit like thick dolphins with short, wide noses. But they're more related to elephants than dolphins.

Sea cows graze on grasses on the sea floor. They're the only marine mammals that are herbivores. Their large lips work like elephant trunks. They grasp and pull food into their mouths. Sea cows move so slow that algae can grow on their backs.

Hyraxes are small, furry animals. They only weigh about as much as a house cat. They're sometimes called rock rabbits. But they're not rabbits at all. They look like rodents. They look nothing like elephants. But they're elephants' closest relatives on land.

Hyraxes live in groups called colonies. They like to sunbathe in the morning. Then they eat quickly. They're always on the lookout for danger. They don't have elephants' size to protect them.

Like elephants, hyraxes are also herbivores. They only eat plants.

CHAPTER 2

HOW ELEPHANTS SAVE THE DAY

Elephants are so huge that it would be impossible for them not to have an impact on the environment. Wildlife species that have a big impact on their whole ecosystem are called **keystone species**. Lots of other species depend on them. Without elephants, their ecosystems would be different places.

Elephants are herbivores. That means they eat only plants. And elephants eat a lot. They spend 12 to 18 hours a day eating. They eat about 250 to 300 pounds (113 to 136 kg) per day. They snack on trees, bark, roots, shrubs, grasses, herbs, and fruit.

All of that eating creates a lot of poop. Elephants can poop up to 15 times a day. As they poop, they drop seeds from the plants they ate. They spread seeds from more than 100 kinds of plants. Elephants plant trees through their poop.

Elephants show affection through touching trunks. They may even do this while they're eating.

DID YOU KNOW?

All three kinds of elephants—African savanna, African forest, and Asian—are **endangered.** African forest elephants are the most endangered. Endangered means they're at risk of extinction. There are fewer elephants every year because of **poaching.** Poaching is the illegal killing of wildlife. Some experts worry that African elephants could disappear within 20 years if more isn't done to protect them.

Forest elephants spread more seeds than any other large animal. Forest ecosystems wouldn't have as many different kinds of plants without them. This keeps forests healthy.

Healthy forests are an important part of fighting **climate change**. Climate change affects all life on Earth. It causes global warming. It causes deadly storms. Forests help clean up the pollution that causes climate change. Trees in forests work together to act as natural air filters.

Elephants are an important part of the **food web**. The food web is how living things depend on each other for food. Dung beetles feed on elephant poop. Then they bury the extra poop underground for their young. This helps fertilize the soil so plants can grow. In turn, the beetles' young are food for other animals like mice and honey badgers.

◀ This African forest elephant is slightly smaller than the African savanna elephant. They live in tropical rainforests in Africa.

Elephant herds make an impact everywhere they go. They trample plants and knock down trees. This helps sunlight reach smaller plants like grasses so they can grow. Other herbivores eat those grasses. Smaller animals use the paths elephants create to travel. Saving elephants helps save whole ecosystems.

Rollers are dung beetles that roll bits of dung into balls. They then roll the balls away and bury them.

Elephant trails usually lead to resources like water or plants elephants like to eat.

SUPER SIDEKICK

DR. PAULA KAHUMBU

Paula Kahumbu grew up in Kenya. She lived in the city and had never seen an elephant up close. But she had heard that elephants were dangerous. One day, she found a book at the library written by a British scientist who studied elephants. She learned about elephants' personalities. She became curious about these extraordinary animals.

Kahumbu kept learning more about elephants. She met other Kenyan women who studied them. That's when she knew she could study elephants, too.

Dr. Kahumbu started a charity called WildlifeDirect. Then she started a movement called Hands Off Our Elephants. It brought people and the government together to protect elephants from being killed for their tusks. It worked to help communities live alongside elephants and want to save them.

WildlifeDirect also works to conserve and legally protect natural areas that are habitat for elephants.

She also started a TV show called *Wildlife Warriors*. It was the first show about African wildlife created by an African person. It inspired people across Africa to care about nature. She's helping train other Africans to make nature films, too.

CHAPTER 3

THE ELEPHANT'S NEMESES

Only one animal is truly dangerous to elephants—humans. The biggest threat to elephants is poaching. More than 20,000 African elephants are killed every year by poachers. Elephant tusks are made of ivory. Ivory has been used by people for hundreds of years. Piano keys were once made of ivory. People carved it into jewelry and figurines. It was seen as a symbol of wealth and power. Millions of elephants were killed for their tusks.

People realized they had to take action before all elephants disappeared. Several countries have made it illegal to buy or sell ivory. Rangers patrol elephant **habitat**. Habitat is the natural home of wildlife. The rangers help protect elephants from poachers.

Even so, too many elephants are still being killed for their tusks. To stop this, governments around the world would need to ban ivory sales. People would need to be educated about why they shouldn't buy it.

Ivory is bone tissue surrounding enamel. Ivory tusks are actually huge teeth!

Elephants and other wildlife are also losing their habitats. Humans are destroying forests. Villages and farms are spreading closer to where the animals live. Many people don't like having elephants around. Elephants will eat crops being grown for people. An elephant raid can destroy a whole field of crops. Then people don't have enough to eat. Farmers lose money.

People are afraid of elephants destroying their farms or communities. This can lead to more elephants being killed.

But more people are learning about **coexistence**. This means living alongside each other in peace. People are finding creative ways to coexist with elephants. Elephants are afraid of bees. Scientists are using beehives to form fences that keep elephants away from villages and farms.

People are also learning that protecting elephants keeps people safe, too. Some rangers are being trained to respond to non-wildlife crimes. They're protecting wildlife and communities at the same time.

DID YOU KNOW?

Rhinoceroses have a lot in common with elephants. They use their horns to dig for water. They also like to roll in mud puddles. That helps make watering holes bigger for other animals to drink from. Rhinos eat a lot of plants and poop out a lot of seeds, too. They help keep their ecosystems healthy.

Some farmers even use firecrackers to scare elephants away from their crops.

HOW TO HELP SAVE THE DAY

Elephants may live on the other side of the planet, but you can still take action to save them.

It's not just ivory that threatens elephants. Other products are responsible for destroying elephant habitat. The forests where Asian elephants live are cut down for wood. They're cleared for coffee and palm oil plantations. Make sure your family buys only elephant-friendly products. Your family can look for fair trade-certified coffee. Wood products are also certified to show they were sustainably harvested. Look for snacks and other foods that don't have palm oil. Choosing elephant-friendly products also helps other wildlife, like tigers and orangutans.

No wild elephants are in the United States. But there are elephants in captivity. Some are still used in circuses and other shows. There are about 160 African elephants in zoos. Many zoos claim that letting people see elephants up close helps save them in the wild. But a lot of

This rainforest in India is being cleared for a plantation, destroying elephant habitat.

zoos have stopped keeping elephants. Experts say that keeping elephants in captivity is cruel. Elephants need a lot of room to roam. They need their families and their freedom.

If you'd like to make a difference, you could choose not to go to places that keep captive elephants. They may stop keeping elephants if they're not making money off them. That could save elephants from suffering. If your local zoo has elephants, you could write them a letter. You could tell them why you won't visit. You could tell them elephants need to be saved in the wild.

ACTIVITY

RAISE MONEY TO SAVE ELEPHANTS

Many organizations are working to save elephants. They're taking action to stop poachers. They're working with communities on coexistence. They're fighting to get ivory sales banned around the world. They're trying to stop people hunting elephants for sport. They're caring for baby elephants whose parents were killed by poachers.

You can help by holding a fundraiser for one of these organizations. And you can help teach others about elephants at the same time.

1 CHOOSE YOUR ORGANIZATION

Search online with an adult for elephant organizations. Learn about what they're doing to save elephants. Pick one doing work you love.

2 DECIDE HOW YOU WANT TO RAISE MONEY

Some kids hold bake sales. Others have sold elephant-themed artwork. You can also throw a Save the Elephants birthday party. Celebrate elephants and ask for donations instead of gifts. You can do a fundraiser on your own with an adult's help. Or you can find a group to work together. It could be a classroom fundraiser, a scout project, or a church group activity.

3 MAKE A LIST OF WHAT YOU NEED FOR YOUR FUNDRAISER

You'll need ingredients and recipes for a bake sale. You'll need craft supplies if you're selling art. You'll need party supplies if you're sharing your birthday celebration with elephants. You'll also need to choose a date and place for your fundraiser. Work with an adult to plan all the details.

4 MAKE SIGNS FOR YOUR FUNDRAISER

Share what you've learned about elephants. Share why elephants need our help. Share information about the organization you're supporting. Make sure your signs are easy to read. You can also make information cards for people to take with them.

5 HOLD YOUR FUNDRAISER

Whether you raise $5 or $500 for elephants, every bit helps. Ask an adult to help you send the money to the organization you chose. Include a note telling them all about your fundraiser. They would likely love to hear how people around the world are helping save elephants.

LEARN MORE

BOOKS:

O'Connell, Caitlin. *The Elephant Scientist.* New York, NY: Clarion Books, 2016.

Ruurs, Margriet. *The Elephant Keeper: Caring for Orphaned Elephants in Zambia.* Toronto, ON: Kids Can Press, 2017.

Markovics, Joyce. *Elephants.* Ann Arbor, MI: Cherry Lake Publishing, 2023.

Viola, Jason. *Science Comics: Elephants: Living Large.* New York, NY: First Second, 2024.

GLOSSARY

climate change (KLIYE-muht CHAYNJ) changes in weather, temperatures, and other natural conditions over time

coexistence (koh-ig-ZIHST-uhns) people and wildlife living alongside each other in peace

ecosystem (EE-koh-sih-stuhm) a place where plants, animals, and the environment rely on each other

endangered (in-DAYN-juhrd) at risk of extinction

extinction (ik-STINK-shuhn) when all of one kind of plant or animal dies

food web (FOOD WEB) a network through which living things depend on each other for food

habitat (HAA-buh-tat) the natural home of plants and animals

herbivore (UHR-buh-vor) an animal that eats only plants

keystone species (KEE-stohn SPEE-sheez) wildlife that have a big impact on their entire ecosystem

matriarch (MAY-tree-ark) the female head of the family

native species (NAY-tiv SPEE-sheez) plants and animals that are a natural part of an ecosystem

plantations (plan-TAY-shuhnz) large farms where one main crop is grown

poaching (POH-ching) hunting animals illegally

INDEX